Ana Isabel Pereira Cardoso
Teresinha M. Gonçalves

Water from a Transdisciplinary Perspective

Ana Isabel Pereira Cardoso
Teresinha M. Gonçalves

Water from a Transdisciplinary Perspective

The inclusion of the subject - the third party

ScienciaScripts

Cover image: www.ingimage.com

This book is a translation from the original published under ISBN 978-3-8417-0281-4.

Publisher:
Sciencia Scripts
is a trademark of
Dodo Books Indian Ocean Ltd. and OmniScriptum S.R.L publishing group

120 High Road, East Finchley, London, N2 9ED, United Kingdom
Str. Armeneasca 28/1, office 1, Chisinau MD-2012, Republic of Moldova, Europe
Printed at: see last page
ISBN: 978-620-7-11994-3

I DEDICATE it to the children,

[...] because all children carry the weight of society bequeathed to them by adults, but they do so with the lightness of renewal and the sense that everything is possible again.

Manuel Jacinto Sarmento

"The main cultural impact of the quantum revolution was certainly the questioning of the contemporary philosophical dogma of the existence of a single level of reality. The quantum revolution played an important role in the birth of a new approach that is at once scientific, cultural, social and spiritual - transdisciplinarity."

Basarab Nicolescu

SUMMARY

1 INTRODUCTION

Knowledge is the key to human action. Its dissemination makes it possible, in a conscious and coherent way, to build a society based on ethical and egalitarian pillars, with a view to the quality of life of its members. The lack of knowledge is causing humanity to go against the grain of its own existence, given the absence of water protection measures and the neglect of this natural resource.

Environmental degradation has brought compromising consequences, even more critically in relation to water pollution, at a high cost to the environment and putting life on the planet at risk. The current scenario (2015) of water scarcity in some regions of Brazil, caused by a period of drought, has brought to light a number of previously unseen factors.

It exposes planning problems, such as the failure to provide reservoirs and the lack of investment in the maintenance of supply networks, which would prevent high levels of wastage by standards. The water crisis also exposes some positive points under transdisciplinary analysis, such as the consumer's realization of their own responsibility and the necessary respect for this vital asset, exposed to the current and future risks of scarcity.

In such a way that the knock-on effect of water shortages puts the need to understand the life cycle of water on the agenda in everyday activities, fostering new knowledge and sustainable know-how in order to reduce environmental impacts.

Classical science seeks answers to all of humanity's questions through rigorous scientific methods and experiments, based on the mathematical logic of measuring, dividing, evaluating and counting. The perfect reproducibility of experimental data makes it possible to repeat the results of research, which makes it possible to predict the future and the final truth.

Based on theories and methods that analyze the object without taking the observer into account, science interprets experiments rigorously in order to determine the product of research as absolute truth. The ideal of classical science is the objective description of phenomena.

When studying the environment, it is possible to see the limitations of classical scientific knowledge, because it is generally made up of infinite varieties of subject/object relationships and determines, in the same proportion, levels of reality, whose events are dynamic and systematic. The product of the experiment is dependent on the observer, the *third party included.*

The term *third included* was conceived by the Romanian Stephane Lupasco, as the element that goes beyond Aristotle's principle of identity, of *being or not being.* As the third element, 'T', is introduced into the experiment and does not reject the principle of identity, it is based on quantum experiments and introduces a new element into the relationship between the subject and the object, i.e. the observer.

The geopolitical characterization of the two cities that are the focus of this research - Criciuma and Urussanga, located in the southern region of Santa Catarina, has a history marked by coal mining, when mining companies used bodies of water as a resource for extraction processes. The rivers that cross the mining routes were used as drainage channels for the polluted water from the mines and the soil was also degraded as pyrite tailings were deposited, resulting in a large environmental liability.

In addition to the historical context of disorganized socio-economic development, which has compromised much of the region's water, the rivers are still being impacted by the acidic water that drains to the surface from deactivated mines. Another aspect of regional development is the growing demand for water from agro-industrial activities, with unsustainable management that continues to compromise the hydrological cycle, water bodies and the environment as a whole. As a result, aquatic communities have disappeared and many bodies of water have lost their natural characteristic of renewal, putting supply at risk.

However, due to its geographical location, the region has a humid subtropical climate, determining a good distribution of rainfall, sufficient to supply the population with water in quantity and quality (TEIXEIRA, 2001, p. 426).

The region where the municipalities surveyed are located is made up of three hydrographic basins, the Ararangua River, the Urussanga River and the Tubarao River, which in one way or another have been compromised by the history of regional development. At the end of the last century, the authorities and politicians of the region's municipalities, concerned about the risk of shortages, began the project to build a dam to meet the continuing demand in the face of a history of environmental degradation and conflicts with agro-industries over supply levels.

Inaugurated in 2006, the 450-hectare Sao Bento river dam currently supplies eight municipalities in the region. Located in the community of Sao Pedro - Sideropolis, SC, on the slopes of the Serra Geral. It has extended the supply by at least 30 years. The catchment results from the damming of the Serrinha and Sao Bento rivers, which belong to the Ararangua River Basin, because "the Criciuma region is critical in terms of the pollution of water resources [...], mainly due to activities linked to coal mining (CASAN, 2014).

Another important factor in the region's history is basic sanitation, which contributes to the multiplicity of environmental problems, as some sewage is still deposited in the city's canals and rivers. On the other hand, in the last six years, Criciuma's municipal administration, together with CASAN - Companhia de Agua e Saneamento (Water and Sanitation Company), has installed a sewage network and treatment plant in some areas of the city, but most of the population (2015) has not yet connected their homes to the public sewage network.

The expansion of industry and agro-industry is another aspect, as they account for the largest share of water consumption and contribute to the contamination of water bodies with agricultural inputs, agro-toxins and animal waste, putting recharge areas and springs in the region at risk, increasing the complexity of environmental problems.

The entire environmental condition of the region is perceived from a transdisciplinary perspective, based on the uncertainties and difficulties related to water today. Faced with the need to minimize the impact of human activity on the environment, through knowledge and, consequently, understanding of the different levels of reality of the subjects involved.

The way in which water is being used exposes the need for measures to protect and sustainably manage bodies of water. In urban centers, the problems are aggravated by poorly structured basic sanitation and the unsustainability of industrial production processes. In rural areas, agricultural production with little structure in terms of environmental preservation and conservation is the focus of the problem.

What is the essence of human work and the social relationships that permeate the daily lives of nations, if it doesn't bring with it quality of life and development for all, in the constant search for greater equality of rights, opportunities and social justice?

Thus the research problem is configured in water as an element of discussion in the context of environmental complexity, dependent on the subject, the *third party included,* and aims to make a substantial contribution with suggestions and analysis of transdisciplinarity as a methodological option for broadening the reductionist view of classical science on the subject of water, through the participation of the subject, listed as the *third party included.*

As well as establishing a comparative analysis of the assumptions of classical science and transdisciplinarity, in order to reflect on the inclusion of the subject, the *third party included, as* an integral element of the process; investigating the perception of the subject, characterizing them as an agent of protection, with their own knowledge and knowledge elaborated in the relationship between water and the life cycle, characterized as the *third party included and*; relating the possibilities for expanding scientific practices through the transdisciplinary approach used for water.

This research consisted of two segments: the literary segment, with a comparative analysis of the presuppositions of classical science and transdisciplinarity, and the *third-party* segment, with interviews with nine citizens and one world personality, the president of NESTLE, a food company, through the analysis of interviews available on the Internet, totaling 10 subjects, who contributed as *third parties.* Different data collection techniques were used for each segment:

The bibliographic and literary research was determined by the collective of actors made up of personalities from classical science and scholars of transdisciplinarity and through the comparative analysis of methodological assumptions. It was carried out in specialized databases on the subject, such as books and bibliometric *software*, such as Scielo and others in Portuguese.

The *Third Included* segment was made up of the collective of citizens from Criciuma and Urussanga, municipalities in the extreme south of Santa Catarina, as they are closest to the research site, using the narrative interview technique, containing five questions, adapted to the five principles of Santos (2008).

Santos (2008) proposed five principles to promote conceptual unity in the construction of knowledge and the contextualization of the knowledge produced. The appropriate hermeneutic technique was used to analyze the data collected through the interviews.

Nine of the ten subjects were chosen because of their involvement in water-related events and the possibility that they had been sensitized to the issue. Another subject was chosen because he was a global personality, a representative of the world's largest water bottler, the food industry (NESTLE), and because he expressed his opinion on water on social networks.

The two municipalities in the extreme south of Santa Catarina, Criciuma and Urussanga, were chosen as the research site because they are part of the region where the researcher lives, and can determine the *level of reality, in the* technical investigation of the data, in the search for knowledge and in the commitment to know and discover.

Based on the exposure of socio-environmental phenomena in the region surveyed, the determination of the sample is intended to establish the relationship between subject/object, i.e. the subject as a *third party included.* As such, the sample was made up of:

Two women, because of their cultural and social condition related to children, family and water. A militant, engaged for years in defending the environment of the Urussanga river, due to the degradation of the soil and water, and the history of unsustainable mineral exploitation, defending sustainability and social promotion. He is a young politician, with a degree in the environmental field and a proponent of social projects for the sustainable use of water in the municipality of Criciuma, SC.

The participation of three students and workers, aged between 19 and 22, was analyzed for the possibility of collaborating with the innovative vision, contributing original ideas and solutions different from the traditional ones. Such participation is also important in terms of interaction with the topic and as conscious and critical citizens, since young people and children are dependent on environmental training and qualification.

A master's degree student; a businessman from the metallurgical industry, with a capitalist and market view of water as a raw material and consumer good; and finally, to make up the total of ten subjects, we used the interview with Mr. Brabeck, president of NESTLE, one of the world's largest food industries, published on *Youtube.com,* defending the need to price water as an alternative to the current related conflicts.

The investigation of the subjects identified as *third parties,* the comparative analysis between the actors of classical science and transdisciplinarity were used to understand the aspects of each segment for the elaboration of knowledge on the subject. This resulted in the achievement of the proposed objective of making a substantial contribution with suggestions and transdisciplinary analysis to broaden the reductionist view of classical science on the subject of water.

2 DISCIPLINARITY, INTERDISCIPLINARITY AND TRANSDISCIPLINARITY

In order to elucidate the subject under study, it was necessary to approach the basic knowledge of the historical and epistemological framework of disciplinarity, interdisciplinarity and transdisciplinarity. It is also important, for the understanding and contextualization of the research, to present the history of the terms applied and relate them to water, in order to point out the possible contributions to the enrichment of the transdisciplinary approach and, finally, to justify the methodological model adopted.

Leff (2008), in his book, *Sustainable Discourses,* points out that science, in the course of human evolution, has created gaps in scientific environmental knowledge. The essentially reductionist worldview focuses on techno-scientific progress and the exploitation of nature, revealing, in recent times, restrictions in presenting solutions to the conflicts resulting from subject/environment interaction.

> It is not enough to recognize the cosmovisions of water and the traditional practices of access, management and use of different cultures. It is not enough to oppose the privatization and commodification of water with a Global Water Convention to protect the ecological heritage of water on the planet and for the future of humanity. There is a greater challenge: to invent *democratic water management* in a globalized world; to reconstruct water as a common good, social practices and processes, associated with all forms of production and consumption, the works of supply, drainage and recycling in which water is interconnected with the processes of production and consumption, and interwoven with the fabric of life (LEFF, 2008, p. 108).

For disciplinary methodology, the subject's relationship with the world is "mapped out", because human thoughts, actions and manifestations are predicted and reduced to linear events, predetermined by methods based on mathematical logic. Thus, scientific production takes place within the sequential and fragmented structure of disciplinary thinking, based on the teachings of *Descartes in the* 17th century, which divides phenomena into parts to be understood and studied.
(CAPRA, 1983, p. 29).

On the other hand, the methodological advance brings interdisciplinarity.

According to Nicolescu (1997), interdisciplinarity allows the specific object of study, which is the domain of a given discipline, to be analyzed by others and thus contribute to, and even modify, the vision and techniques of the research, providing new results and a new perception of it. Interdisciplinarity is the matrix in which each discipline researches and analyzes the specific object of its area of knowledge, crossing borders.

The concepts of interdisciplinarity emerged in 1961 in the human sciences. "[...] from a project developed by Georges Gusdorf[1] for UNESCO and was based on a new explanation of the nature of reality and knowledge, from a new scientific paradigm focused on the uniqueness of knowledge and overcoming its fragmented or disconnected nature." (MORAES, 2008, p. 62).

In the last century, with the proposal of transdisciplinarity as a methodology, it has its starting point in disciplinarity. Transdisciplinarity analyzes the object to the extent that it perceives the existing levels of reality and the complexity of the subject's thinking, but it goes beyond interdisciplinarity and is dependent on it (Nicolescu, 1999).

The search is not complete, it is dynamic and links the product of the research to the researcher, their knowledge, thoughts, reflections and feelings. Not just through rational capacities, but through all faculties, history and personal culture. This includes analyzing the object of research from the point of view of various types of knowledge: popular, philosophical, religious and scientific (MORIN, 2011, p. 11).

The term transdisciplinarity was coined by Jean Piaget in 1970 in Nice, France, at the *First International Seminar on Pluri- and Interdisciplinarity,* stating that this stage should be followed by a transdisciplinary one. Later, at the Colloquium, "Science at the Frontiers of Knowledge", held in Venice, Italy, in 1986, transdisciplinarity was suggested as a scientific method for human understanding and appropriation of the various types of knowledge, since the accelerated and unprecedented production of knowledge and society's inability to make use of it has reached a level where it puts human existence and that of planet Earth at risk (Venice

[1] GEORGES GUSDORF (1912), a Frenchman, determined that there was a need for "specialists of non-specialties", with a scientific and humanistic culture beyond the narrow field they studied.

Declaration. UNESCO, 1986).

In the "Manifesto of Transdisciplinarity" Bassab Nicolescu (1999) states that: "Tomorrow will be too late." Because the ultimate goal of science is uncertain, each area of knowledge has exploited knowledge with its own knowledge, techniques and interests, more specifically the environment and water. Sometimes they fail to realize the inter- and transdisciplinary nature necessary to protect this vital element for life.

The purpose of science today should be to improve and guarantee the conditions of life on Earth, be it animal, vegetable or human, for current and future generations. [...] the product of a blind but triumphant "technoscience" that only obeys the implacable logic of efficiency for efficiency's sake (NICOLESCU, 1999).

> The major environmental issues of our time (namely global warming, lethal air and water pollution, food insecurity, the clear depletion of natural resources, criminal deforestation and widespread soil degradation, to name but a few) must be understood as natural, social and economic issues, which is why they can only be addressed through an integrated, objective, strongly empirical and, in a word, systematic approach (FREITAS. 2012, p. 31).

The fragmented conception of nature, the increasingly reductionist and disciplinary view, dominated by rational thinking, considered the only acceptable way of producing knowledge, has contributed to the environmental crisis. It has a huge influence on most sciences as a logical basis for dealing with the environment, which is studied in parts and exploited by different interest groups with profoundly anti-ecological attitudes, hindering the perception and transdisciplinary conception of ecosystems. "The environment, like knowledge and understanding, changes over time. Logic can therefore only have an *empirical foundation* (NICOLESCU, 1999).

Transdisciplinarity uses concepts perceived in the interrelationship, so knowledge originates in experience. Perceiving the world as a system of inseparable components in permanent dynamic interaction, with the subject being an integral part of the system, is the basis of knowing, and knowing is impossible without experiencing.

Because it establishes a relationship between knowledge, transdisciplinarity, as a scientific approach, can be considered something new, a neologism, when it is perceived in its totality and in the scope of the various areas of knowledge. Thus, the production of knowledge occurs with the participation of the subject, through experimentation, expressed in the various ways in which the perception of the senses is interpreted.

Scientific experiments and transdisciplinary experiences are unique. With each repetition, the results change, modified by the environment, the subject or the object which, in turn, are modified by the experiment.

> These disjunctive/repulsive terms, while canceling each other out, are at the same time inseparable. The part of reality hidden by the object refers back to the subject, the part of reality hidden by the subject refers back to the object. What's more, there is only an object in relation to a subject (who observes, isolates, defines, thinks) and there is only a subject in relation to an objective environment (which allows them to recognize themselves, define themselves, think, etc., but also to *exist)* (MORIN, 2011, p. 41).

Science is dependent on mathematical logic, and knowledge, the result of experiments using disciplinary methods and techniques, is restricted in translating the different levels of reality exposed. Unlike transdisciplinary logic, which is rich in paradoxical, indeterminate and imprecise situations and is based on diverse subject/object interactions.

Consequently, the experiment acts in the zones of indefiniteness and domains of ignorance, and by occupying these spaces, it can generate new disciplines, or remain as free zones, circulating between the disciplinary interstices, in such a way that transdisciplinarity is left with the movement, the indefinite and the inconclusive of research knowledge.

3 SUBJECTS AND THE THIRD INCLUDED

The transdisciplinary approach adopted as a methodology aims to stimulate an understanding of current environmental complexity, in order to optimize social projects and actions with the involvement of the subject, the *third party included,* in the search for participation and responsibility on the part of everyone.

The appropriation of nature, combined with uncontrolled environmental exploitation, modifies ecosystems, watercourses and the natural habitats of plants and animals. All human activity produces waste, polluting and interfering with the natural hydrological cycle. In addition to being costly, it makes the process of making water drinkable more difficult and longer.

The growing demand for water and economic relations with it require a perception beyond its use, with the involvement of the subject, recognizing it as humanity's heritage, essential to the life chain and dependent on care. The Charter of Principles for the Transdisciplinary Study of Water/CET - Agua, states that: "Transdisciplinarity [...] recognizes the existence of different levels of reality governed by different logics; it refers to the different levels of perception of the subject and the different levels of reality of the object." (CET - Agua, 2011).

Water washes everything away, seems to have dominated the subject's thinking, because traditionally sewage, solid waste and industrial effluents are often dumped into the environment, which is unable to degrade and regain its natural balance, decimating aquatic communities and the entire chain of ecological balance.

Socio-economic development has historically compromised water, as it is a basic element of any project or action and impacts, greater or lesser, are always present. Demand from industries and agro-industries depends on sustainable management to minimize damage to the environment as a whole. Data from the National Water Agency (ANA) (2013, p. 89) shows that in 2010, demand in the agricultural sector (irrigation/animal production) and industry accounted for 88% of water consumption, while rural and urban human consumption accounted for 12%.

Agudo et al (2014) states that water as a commodity, a natural resource, promotes the accumulation of capital by large corporations, especially the agricultural sector: "it uses water as added value in commodities, increasing foreign exchange on the international market. [...] Meanwhile, the 'culprit' for the water crisis is the population." (AGUDO et al, 2014).

Otherwise, the profits from appropriation do not, for the most part, go back into the water, guaranteeing care for the hydrological cycle of natural renewal. The myth of water as an infinite resource prevails, and not as a vital asset for all species of life dependent on the hydrological cycle. For the capitalist world view, water is a "market commodity". Many companies use bodies of water to their advantage, based on the law of supply and demand, with the aim of making a profit. "The move towards sustainability implies the gradual deconstruction of the prevailing anti-ecological and entropizing economy." (LEFF, 2008, p. 38).

The transdisciplinary perception in the logic of the *third party included,* determines the impossibility of involving the subject without taking into account their own reality, because the simple human manifestation and use, commits them and involves them in socio-environmental processes.

In order to understand the concept of the *included third, it is* necessary to know the epistemological relationship between the concept and the term. The Romanian Stephane Lupasco, born in 1900, idealized the logic of the *included third,* the element that goes beyond the duality foreseen by the principle of identity of *being or not* being. He introduced the third element 'T', and generalized logic to include physics, epistemology and consciousness. However, Lupasco did not reject the principle of identity, but only introduced a new element to human relations with nature.

The famous state T ('T' of the *third included),* according to Nicolescu, (2001, p. 110), appears for the first time on page 10 of Lupasco's book *Le Principe d 'antagonisme,* defined as a state "[...] neither actual nor potential" and the *third included,* underlying the principle of antagonism, being the *unifying third,* it unites being and *non-being.* To better understand the third element, it is necessary to understand the *levels of reality, a* term proposed by Nicolescu in 1985.

> [...] I suddenly realized that this notion also provided a simple and clear explanation for the inclusion of the third party. [...] Reality is not only a social construction, the consensus of a community, an intersubjective agreement. It is also a *trans-subjective* dimension, to the

extent that a simple experimental fact can ruin the most beautiful scientific theory (NICOLESCU, 2001, p. 120).

The individual, if committed to the environment and life on Earth, realizes that human survival and manifestation produce situations that risk the integrity of the hydrological cycle, which compromises and hinders the process of making it drinkable. Thus, it becomes an agent of environmental protection, because it understands that water is essential to the life cycle and is everyone's right and responsibility. The demand for water is growing and the complexity of human relations with it requires greater involvement beyond simple use.

The transdisciplinary view of environmental conflicts perceived in an integral way becomes an instrument for analysis and reflection, reworking new sustainable practices, with a gradual change in attitude, in the interdependence of the subject with nature, a participatory and co-responsible agent, contributing ideas and solutions, in preservation and conservation for their own benefit.

The transdisciplinary methodology as a research approach is characterized by the production of knowledge through broad, flexible paradigms that are more suited to the historical-cultural subject. Determining the complexity of their relationship with water and their own level of reality.

The multifaceted nature of the exploitation of natural resources makes it necessary to broaden the scientific model and propose new ways of using water. Water, manifested in all things and the most common element on Earth, is as dependent on the subject as the subject is on it, in order to preserve the hydrological cycle necessary for the very renewal of life.

4 THE TRANSDISCIPLINARITY OF WATER

Every science requires presuppositions, whether it's the traditional so-called classics or the new ones. Transdisciplinarity is based on practices, experiments and procedures in the production of knowledge based on new logics of human interrelationship with nature. The subject's meanings and values, and reality itself, engage him or her in a broad way with today's political, economic and environmental issues. "A transdisciplinary attitude is open to the new, not confusing the different manifestations with the 'real'" (MORAES, 2008, p. 24).

The promotion and expansion of access to knowledge has been the driving force behind development and quality of life. Contrary to the history of past civilizations, which suffered as a result of the prohibition of access to information, under penalty of death, today we are faced with an enormous production of knowledge that is available through various means and sources of knowledge. The challenge then turns to the ability to understand, discern and use what is available.

The main means of access to scientific knowledge is still the school and it is up to us to play our part in training people in the management and conscious use of water, promoting a broader vision through a transdisciplinary approach with research, the production of knowledge and concepts, involving everyone.

Classical science determines interpretations and experiments with theories and methods, with the product of research constituting the "absolute truth". However, analysis and logical reasoning are incapable of immediately explaining environmental and social phenomena, by isolating a certain group of phenomena and building a model, because they disregard the totality, without perceiving the levels of reality of the subject involved.

Classical scientific rigor determines the separation of the *self and the object,* but scientific evolution, supported by quantum theory, has modified this rigor, demonstrating that in subatomic experiments the observer is an element of participation in the experiment. The properties of any subatomic particle can only be understood in interaction with the observer. Thus, the classical ideal of the objective description of nature loses its validity, because experiments are influenced by and influence the observer. Capra (1983, p. 110) quotes: "Natural science", says Heisenberg[2] , "does not simply describe and explain nature; it is part of the interaction between nature and ourselves".

In this way, transdisciplinarity, based on the logic of the third party included, does not deny disciplinarity and transdisciplinary experiences. It may seem at odds with the scientific methods of classical science, but it is dependent on disciplinary, multidisciplinary and interdisciplinary practices. It relies on these, as a perceptive action on the object of study, analyzing, reflecting and reworking knowledge in the present world. Nicolescu (2000, p.139) defines the practice of transdisciplinarity as: "[...] the embodiment of each action, of the transdisciplinary methodology, through a set of methods adapted to each specific situation".

This incarnation, Nicolescu (2000) adds, seems incomprehensible at first, because it is difficult to perceive something that has never been seen before. It may also seem unattainable in its fullness, but a gradual change in attitude is necessary, as is the perception that gradually impregnates new generations with notions of ecology and the scope of this concept, in the interdependence of the subject and nature.

The transdisciplinary experiment depends on the reality of the subjects involved, their own knowledge developed according to their perception of the world and the aspects inherent in the object of study. "Knowledge is never definitive, but a product of humanity, always linked to historical circumstances, which are as dynamic as the individuals who experience and project it". (SANTOS, 2008, p. 76).

In this way, classical science has sought the final truth, the correct answer to humanity's challenges. This final, conclusive truth is obtained by scientific methods; it is called determinism. We can follow the trajectory of science over time, especially from the end of the last century to the present day, culminating in the production and accumulation of decontextualized knowledge, with fragmented research. "The equations of classical physics are such that [...] we can predict their positions and velocities at any other moment in time" (NICOLESCU, 1999).

[2] HEISENBERG, (1901), German theoretical physicist who received the Nobel Prize for Physics, (1932), creator of quantum mechanics.

Every event observed through classical science has a local origin and causality. The subject, observer or researcher, the agent of the interpretation of phenomena, has no influence on the experiment and must remain exempt from it. The evolution of phenomena takes place within the linear parameters of time and space, on a single level of reality, as in Hegel's triad: Thesis - Antithesis - Synthesis, and in this way the reconciliation of opposites is not possible. "In the logic of the third included, the opposites are rather *contradictory:* the tension between the contradictory promotes a broader unity that includes them" (NICOLESCU, 1999).

The transdisciplinary approach, paradoxically, is the main obstacle to broadening the classical view in terms of understanding concepts and processes. Based on subjective aspects and the very characteristics of relativity, it does not offer a *firm solution* in terms of methodological determination and the stages of research, requiring resilience on the part of the participants.

Nicolescu (1999) draws a parallel between transdisciplinarity and the postulates of classical science, when he analyzes the fact that not all cutting-edge sciences can comply with the three postulates of such rigorous mathematical formalization of science: a) the existence of universal, mathematical laws; b) the discovery of laws through scientific experimentation and, c) the perfect reproducibility of experimental data.

Nicolescu then formulates the three pillars of transdisciplinarity: 01. Complexity; 02. The logic of the third included and; 03. The levels of Reality - determining the transdisciplinary methodology.

> In other words, there are degrees of disciplinarity, proportional to the greater or lesser satisfaction of the three methodological postulates of modern science. In the same way, the greater or lesser satisfaction of the three methodological pillars of transdisciplinary research generates different degrees of transdisciplinarity. Transdisciplinary research corresponding to a certain degree of transdisciplinarity will be closer to multidisciplinarity (as in the case of ethics); to another degree, it will be closer to interdisciplinarity (as in the case of epistemology); and to yet another degree, it will be closer to disciplinarity (NICOLESCU, 1999).

The three pillars of transdisciplinarity in the text below are reinterpreted in their aspects and concepts to facilitate a transdisciplinary understanding of water.

The first pillar of transdisciplinarity is *complexity, reflected* in the extent of the interconnected relationships between subject/object, i.e. subject/water. Perceived in the diverse knowledge produced about water in socio-environmental conflicts, in the inalienable right and responsibility for this common and intergenerational good, it reflects the complexity of the subject.

> The complexity of science is first and foremost the complexity of equations and models. It is therefore a product of our minds, which are complex by their very nature. However, this complexity is the reflected image of the complexity of experimental data, which accumulates endlessly. It is therefore also in the nature of things (NICOLESCU, 1999, p. 19).

The French sociologist Edgar Morin (2011), observing the frequent transformations of information, realized the urgent need to overcome the boundaries between disciplines and suggested complexity in order to develop new conceptions of knowledge and thus extend the trend towards specialization and fragmentation of knowledge.

The second pillar, the *third included,* is determined by the subject, positioned at the top of rationality as a thinking being, acting critically in the exercise of citizenship. Sensitized to perceive their own condition as a *third party,* crucially involved in the interdependence of their own existence with water and committed to the natural water cycle. "Suffice it to say that the included third party induces the *discontinuity* of space and time." (NICOLESCU, 2001, p.119).

The transdisciplinary perception of the relationships established with water, in the face of the logic of the *third party included,* compromises the other two pillars, *complexity and the levels of reality,* because it is impossible to involve the subject without taking into account their own reality in the process. In this way, the dialogic nature of the relationships gives rise to data of complexity, the more it is used, the more it is committed to and involved in socio-environmental processes.

The *Levels of Reality, the* third and final pillar, can be understood as the subjectivity of each individual, subject, society, culture or phenomenon, with a complex structural composition, in the simultaneous occurrence and influence of what we are. This pillar, related to water, is represented in the wide range of social, cultural and economic relationships with water and in the satisfaction of the vast majority of living beings' biological needs. The different realities can also influence solutions and decisions in water-dependent production processes, as well as the growing demand and possible disturbances in the hydrological cycle.

The *Levels of Reality* - arise in the knowledge of nature, society and ourselves, as a complex structure, each level is what it is because all the levels exist simultaneously, "[...]. These levels of perception allow for an increasingly general, unifying, all-encompassing view of Reality, without ever completely exhausting it. As in the case of the levels of Reality, the coherence of the levels of perception presupposes a zone of non-resistance to perception" (NICOLESCU, 2000, p. 132).

Environmental sciences really come into their own when research takes into account the participation of the subject, because the environment is analyzed in relation to the subject and institutionalizes the *third party included.* The subject, the subjective agent, loaded with meaning by the very history and culture of human existence, can influence and be influenced by water.

Santos (2008) also proposes five principles integrated into the theory of complexity and transdisciplinarity: the Holographic Principle, the Principle of Transdisciplinarity, the Principle of Complementarity of Opposites, the Principle of Uncertainty and the Principle of Autopoiesis.

These principles used in the interpretation of relationships with water have been adapted in the text below, useful in the elaboration of the research narrative interview and in the descriptive critical analysis of the data obtained.

The holographic principle is based on the concept that the whole is in the part and the part is in the whole, and reinterpreted for the subject in question, the principle prioritizes an integral understanding of phenomena. When analyzing the complexity of water, it is necessary to perceive the whole, ecology as a scientific basis and, when deepening the analysis to understand environmental issues, it is necessary to adopt a more comprehensive approach, in the overall view of nature, genesis, evolution and current impasses.

Contextualization is necessary to explain and make sense of isolated phenomena. The parts can only be understood from their interaction with the dynamics of the whole, highlighting the multiplicity of interacting elements which, to the extent of their integration, reveal the existence of different levels of reality, opening up the possibility of new visions of the same reality (SANTOS, 2008, p. 74).

CET - Agua states that the transdisciplinary view of water includes and goes beyond a linear understanding. "It presupposes a comprehensive look at its many aspects and its powerful influence on everything that exists."

Understanding water involves understanding the whole, the commitment of the subject and the levels of reality. Today, humanity is already critically aware of urban conflicts, the tax burden, the energy crisis, environmental disasters and the problem of water as part of the same crisis, which is based on an outdated perception of the world. The global events that interconnect different areas of knowledge bring out the need for an ecological perspective, in the holographic perception of phenomena.

In this way, transdisciplinarity makes it possible to raise the awareness and greater involvement of the subject and encourages a contextualized perception of the world, whose approach to knowledge seeks, in a holographic way, to give broad meaning to human relations with water.

According to Santos (2008), transdisciplinarity and complexity are theories that suggest overcoming the dichotomous thinking "[...] of dualities (subject-object, part-whole, reason-mode, etc.) that came from the vision disseminated by Descartes, stimulating a way of thinking marked by articulation".This principle reaffirms the triad of the *third included, which* promotes the reconciliation of pairs, in the simultaneity of subject/object, part/whole and reason/meaning.

Transdisciplinarity refers to what is at once between disciplines, across disciplines and beyond any discipline. "Its objective is the understanding of the present world, for which one of the imperatives is the unity of knowledge." (NICOLESCU, 1999).

The transdisciplinary perspective is fundamental to promoting greater involvement of people in resolving conflicts generated by water. It recognizes that all phenomena are complex and water, a phenomenon with social, political and economic dimensions, links history, culture, religion and other knowledge. The

Transdisciplinary Study of Water refers to:

Transdisciplinarity is understood in this context as an expanded vision of the world that inspires and supports absolute respect for differences; the dialogue of knowledge between science, art, culture and spiritual experience; an open, multi-referential, multidimensional and transcultural rationality; rigor that takes all data into account as a way of legitimizing synthesis; recognition of the Earth as home to all; and recognition of the human being as a transnational being, one with all life and its environment (CET-Water, 2011).

In September 1927, in *Como,* Italy, during the International Congress of Physics, Niels Bohr presented the formulation of *complementarity to* the public, according to Holton (1984, p. 49): "He (Bohr) was referring to a profound and persistent difference between the classical description and the quantum description of physical phenomena." By proposing an element that is antagonistic, but not opposed, to any phenomenon and with the theory of relativity and quantum mechanics, the proposition of complementarity came to be used to describe the phenomena of nature.

In China, according to Capra (1983), the polar relationship of opposites is at the basis of *Taoist* thought and considers all changes in nature to be manifestations of the dynamic interaction between polarities of *yin* and *yang* opposites, leading *Taoists* to believe that any pair of opposites constitutes the polar relationship in which each of the poles is dynamically linked to the other.

When we look at the opposites in the historical, social, cultural and scientific context, in particular, we see a certain imbalance: in the prevalence of classical science to the detriment of spirituality; of rational thought and intuitive folk wisdom; of the exhaustive and indiscriminate use of nature, without realizing and valuing the need for preservation and conservation; of absolute truth to explain natural phenomena, without understanding the need for relative truth and the provisional answers of environmental and ecological knowledge. "Buddhists talk about 'relative' and 'absolute' knowledge, or about 'conditional truth' and 'truth'" (CAPRA, 1993, p. 28).

In 1938, at the Congress of Anthropology and Ethnography in Copenhagen, Bohr proposed that the human sciences use the principle of complementarity (HOLTON, 1984, p.52). At the same time, in 1951, Sthephane Lupasco proposed the logic of antagonism in The *Principle of Antagonism,* formulating the 'fundamental postulate of the dynamic logic of contradiction' (NICOLESCU, 2001, p. 110).

Nicolescu cites Lupasco's proposal:

Every phenomenon, element or logical event, and therefore the judgment that thinks it, the proposition that expresses it, the sign that symbolizes it - *and,* for example, there must always be associated, structurally and functionally, an anti-phenomenon or anti-element, or anti-logical event, and therefore a judgment, a proposition, a contradictory sign - *ndo-e* [...] (NICOLESCU, 2001, p. 110).

Lupasco always stresses that this is not opposition, but contradiction, together with the *actualization of* the phenomenon and the *potentialization of* its opposite. "[...] *e* can only be potentiated by the actualization of *ndo-e,* but not disappear. Likewise, *ndo-e can only be potentiated* by the actualization of *e, but* not disappear."
(NICOLESCU, 2001, p. 110).

Actualization and potentialization depend on dynamics and the existence of a state of rigorous, exact equilibrium between the poles of contradiction (NICOLESCU, 2009, p. 2).

The manifestation of any phenomenon is equivalent to a certain actualization, a tendency towards identity, but this same manifestation implies a containment, a potentialization of everything that this phenomenon is not, in other words, of non-identity. Potentiation is not an annihilation, a disappearance, but rather a kind of memorization of what has not yet manifested (NICOLESCU, 2009, p. 2).

Actualization and potentiation are perceived when they contribute to the dynamics of the complementarity of opposites; this is possible with water today, whose *actualization of* scarcity *potentiates* the contribution to sustainable environmental attitudes with improvements in abstraction, distribution and quality use, among others. In times of crisis [actualization], it is possible to perceive reality and thus stimulate human intellectual production [potentialization], which is possible in the face of this reality.

Science, by means of methods and techniques, defines the cause and origin of events, determines the effects and thus predicts the future, with absolute certainty of detail. However, transdisciplinarity determines relative truth in the interaction between the subject and the object. The principle of uncertainty is related to the inability to produce absolute truth to describe natural phenomena, because descriptions of space and time and classical causality do not apply. "The principle of uncertainty legitimizes what has always happened in everyday life and accompanies man from birth to death" (SANTOS, 2008, p. 79).

The current condition of water shows the limitation of the mechanistic foundation in determining certainties and absolute truth, to trends and nature analyzed based on events, interaction and experimentation. Environmental experiments tend towards uncertainty and phenomena are interpreted in the universe of relationships. "But complexity is not reduced to uncertainty, *it is uncertainty within organized systems"* (MORIN, 2011, p. 35).

The uncertainty principle was proposed almost a century ago and challenged scientific determinism, expressed by Heisenberg in a set of mathematical relationships that determine the application of classical concepts to atomic phenomena. Concepts and classifications can no longer be presented as absolute truths. Experimentation is what has made it possible to explain events and determine data or justify phenomena.

> While since Hegel and Nietzsche no-verdad has appeared on the horizon of truth, science has been discovering the flaws in the scientific project of modernity, from the irrationality of the unconscious (Freud) and the principle of indeterminacy (Heisenberg), to deterministic chaos, the encounter with the arrow of time and dissipative structures.
>
> (Prigogine). Complexity thinking and environmental knowledge incorporate uncertainty, indeterminacy, irrationality and possibility into the field of knowledge (LEFF, 2008, p. 20).

Transdisciplinarity is based on interrelationships and the constant production and renewal of knowledge, resulting in relative and transitory truths. These aspects point to new investigations into scientific inaccuracy, or the subordinating of theory to its observer. "The more one descriptive aspect is emphasized, the more the other becomes uncertain, and the precise relationship between the two is given by the principle of uncertainty" (CAPRA, 1982, p. 63).

However, traditional science expects knowledge to be objective, demonstrative and without taking into account historical and human interpretation, being able to predict future events deduced from observed reality.

The Principle of Autopoiesis is the term coined in the 1970s by Humberto Maturana and Francisco Varela[3] to define the capacity of living beings to self-regulate and self-produce. "Or, in other words, corporeality, which is when the autopoiesis of the living system actually occurs, is the condition of possibility of the living system, but the mode of its constitution and realization continues and in itself is continuously modulated by the flow of the living system's life in the domain in which it functions as a totality" (MATURANA, 2001, p. 177).

Autopoiesis is the ability of living systems to self-organize, to produce themselves and their own identity, distinguishing themselves in the environment. Nature is made up of interrelated systems organized for self-sufficiency, producing and recycling their own components. [...] two or more organisms interacting repeatedly generate a social coupling in which they are reciprocally involved in the realization of their respective autopoieses (MATURANA, 1995, p. 230).

Because of its cycle, water is an element present in circularity, in each system and between systems, contributing to autopoiesis.

> In the appropriation of knowledge, the subject, by mobilizing their individual characteristics, does not maintain the reliability of the information. This is why it is said that all knowledge is a reconstruction of knowledge. It is a self-organizing activity that produces differences due to the diversity of the environment, human relationships, the genetic load of each individual and their own life history (SANTOS, 2008, p. 80).

Autopoiesis is exposed in society's relationships with the various purposes for which water is used, determining sustainable attitudes for the protection and natural regeneration of living systems, which depend on water as a vital element for self-organization. It is in this that they are simultaneously realized and specified. However, autopoiesis is only possible if the hydrological life cycle is maintained.

The concept of autopoiesis encompasses the phenomenon of self-transformation and self-transcendence, expressed progressively in the processes of interaction with the world. They are present in the dynamic interaction and evolutionary mechanisms of individuals, species, ecosystems, societies and cultures.

Autopoiesis, dependent on water, is confirmed in creative self-overcoming, in

[3] MATURANA (1928) critic of mathematical realism and the biology of knowing, proponents of systems thinking, radical constructivism and creator of the theory of autopoiesis with Francisco Varela (1946).

learning through experience, in genetic mutation due to the physiological need for survival and in the perpetuation of the species.

The five principles proposed by Santos (2008) to reduce the distance between knowledge and useful action, for human development and adapted to water, become tautochronous when analyzed from an ecological perspective - *Autopoiesis,* characteristic of living beings, maintains interactions, triggers determined changes in its own structure and in its function to adapt to the environment. It thus self-produces and self-regulates systems, with *holographic* translations, consequently the principle of the *complementarity of opposites is* present in the constant adaptive changes that characterize *uncertainty,* in the diverse provocations and *complexity of* challenges and stimuli from the environment.

The transdisciplinary proposition can only be sustained when it is based on the inter-relationship or trans-relationship between subject and object, with the ethical objective of developing humanity as a whole. It is necessary to develop resilience and tolerance, consciential attributes based on a transdisciplinary, transcultural, transreligious and transcontinental attitude.

5 LITERARY SEGMENT

In order to facilitate understanding of the relationship between the paradigms of these sciences, the classical and the transdisciplinary, and considering the complexity of the facts, in the complementary approaches in comparative table 1, this chapter analyzes and determines the possibilities for expanding scientific practices and seeks to promote the discovery of regularities and divergences, models and typologies, continuities and discontinuities, similarities and differences, concordances and contradictions between them, for the production of transdisciplinary knowledge.

The first characteristic aspect of the comparison between the assumptions is the inclusion of the third term 'T', the subject who observes and diversifies scientific production beyond the simple analysis of the phenomenon. The approach based on experimentation determines the possibility of the observer influencing the experiment, the experiment influencing the observer and, consequently, influencing the environment, which influences both.

The Cartesian characteristic of fragmenting and reducing the whole in order to be understood leads to the decontextualization of knowledge, widening the gap between the knowledge produced and its effective use for the promotion of life and development. In this way, the critical and participatory responsibility of the *third party* [the subject] is the outstanding transdisciplinary element in the comparison framework. This commitment promotes benefits for the environment and the subject themselves, as well as opening up possibilities for greater use of scientific production in improving the quality of life.

Based on the comparison between simplicity and complexity in the second aspect of table 1, it is possible, through complexity and a systematic view of nature, to produce contextualized knowledge with greater usability. The production of transdisciplinary knowledge makes it possible to include the subjective realities of the subjects involved, in order to broaden the characteristic of discipline, based on mathematical logic methods of quantifying and classifying.

The other aspect, local causality, is determined by the classical principle of separability, where the subject is analyzed separately from the object, they do not interact, and natural phenomena are analyzed only by their local causality. The transdisciplinary principle of Inseparability, on the other hand, refers to the characteristic of natural phenomena.

Subject and object, even if separated at a distance, continue to interact and phenomena can have their global causality, thus committing the subject to the product of the manifestation itself.

Finally, transdisciplinary experiments based on the triad of the third included have the characteristic of simultaneity. Transdisciplinary processes are diffuse, they coexist and the production and use of results is simultaneous. This confirms the broadening of the classical view of nature, since the characteristic of environmental experiments is that projects, actions and results coexist in order not to compromise the conditions and promotion of life, whatever it may be.

Table 01 below presents a detailed comparative analysis of the different aspects in the formulation of environmental scientific knowledge, techniques or approaches, reaffirming the possibilities of expanding the predominant scientific practices:

Table 1 - Comparison of the principles of Classical Science and Transdisciplinarity

CLASSICAL SCIENCE Galileo[4] , Hegel[5] and Newton[6]	TRANSDISCIPLINARITY Nicolescu[7] , Lupasco[8] and Morin[9]
Principle of Identity: Axiom of identity: A and A; Axiom of non-contradiction: A is not and not - A; Axiom of the excluded third, there is no third term T e, the only level of Reality.	Logic of the third included: Pairs of mutually exclusive contradictories (A and not - A); Third included: third term T, and at the same time, (A and not - A) and Levels of realities.

[4] Galileo (1564) A key figure in the classical scientific revolution.

[5] HEGEL (1770) The mind manifests itself as a set of contradictions and oppositions, which ultimately integrate and unite, without eliminating either pole or reducing one to the other.

[6] NEWTON (1642) the function of science was to discover universal laws and to state them precisely and rationally.

[7] NICOLESCU (1942) President and founder of CIRET, proposed the three pillars of Transdisciplinarity.

[8] LUPASCO (1900) Founding member of CIRET, proposed the third term T, the Third Included.

[9] MORIN (1921) Contemporary thinker and one of the main theorists of Complexity.

Simplicity: Discipline; Reductionism, accumulation of knowledge and fragmentation of the whole.	Complexity: The whole is in the part and the part is in the whole. The whole influences the parts and the parts influence the whole.
Determinism: Determines the final truth by scientific methods of measuring, dividing and evaluating, counting.	Uncertainty: Determines relative truth; The transience of truth and The method lies in the subject/object
Objectivity: Subject does not influence the experiment and Separation of subject from object	Subjectivity: Impossible to separate the subject/object and Influence of the observer on the experiment and vice versa.
Single (objective) reality: Objectivity, one and the same level of reality can only provoke antagonistic oppositions.	Levels of Reality (subjective): Subjectivity, at least as many realities as there are people, and Complementarity of opposites.
Separability: If subject and object interact at a given moment and then move apart, they obviously interact less and less - Local causality.	Inseparability: If subject and object interact at a given moment and then move apart, they continue to interact. The observer influences the observed object - Global causality.
Irreversibility: Hegelian triad Thesis-antithesis-synthesis. The three terms of the Hegelian triad follow one another in time. The Hegelian triad is therefore incapable of reconciling opposites.	Reversibility: Triad of the third included: It is capable of promoting the reconciliation of opposites. The tension between the contradictory promotes the broader unity that includes them. Thesis-antithesis-synthesis. Simultaneously coexist.

Source: adapted from Nicolescu (1999).

In summary, Table 2 reinterprets the three transdisciplinary pillars and relates them to water. In the disciplinary and objective analysis, water is an object of consumption, and is conceived as a commodity, a chemical element, or even an infinite resource. The subject does not interact with it and takes little part in environmental decisions. The perceived reality is physical and immediate, of local causality.

In the transdisciplinary perception of antagonism, the physical reality is immediate, water is perceived as more than a chemical element and an object of consumption, it is the vital good, essential for the maintenance of life, whose renewal depends on care.

The purpose of the comparison is not to determine the validity of the method, but rather to broaden the scientific vision, promoting a better understanding of the phenomena. The evolution of the sciences has been marked by going beyond the boundaries of rational thought. It integrates the transdisciplinary perspective into comparative studies, as it helps in the critical and reflective analysis of the phenomenon and the perception of multiple contaminations and crossings, considered in the transdisciplinary perspective as the axis of relations with other knowledge.

Even if the phenomena go beyond the explanations of traditional science and transdisciplinarity is suggested as a methodological approach, it is necessary to recognize the need to apply classical methods to describe the results of the experiment.

It is therefore possible, with the parallel drawn in table 1, to reinterpret the approaches and use them for the critical analysis of water in the face of classical science and transdisciplinarity, insofar as it compromises the subject, values the various realities of water, and realizes the complexity of the phenomenon.

The proposal in box 02 reinterprets the three transdisciplinary pillars and relates them to water. It compares and suggests new tools and methods for expanding the frontiers of science.

Table 2 - Comparison of the principles of Classical Science and the three pillars of the Transdisciplinarity, reinterpreted in the light of the theme

Classical Science	Transdisciplinarity
Simplicity (identity): Water is an object of consumption, a commodity, its use determined by demand, and it must be priced like any food.	Complexity (identity): Water is perceived as more than just a physical element. It is a vital common good whose renewal depends on care and use determined by sustainability.
Excluded third party: The subject contributes little to solving problems, which is the responsibility of the authorities, specialists and experts in the field.	Third party included: The subject actively contributes to all stages of the search for solutions, influences and is influenced by it.
Single reality: Water is analyzed in terms of physical and immediate reality, with local causality as the only reality of perception.	Levels of reality: Water is perceived at different levels of realities, based on experimentation, procedures, and the complexity of subject/object relationships, global causality

Source: prepared by the researcher (2015).

The interpretation of the pillars of transdisciplinarity, adapted to water, can promote appropriation by the subject, the *third party included,* as a vital common good.

In his book, *Transdisciplinaridad y formation universitaria,* Nicolescu (2014)

states that the existence of levels of reality has been affirmed by different traditions and civilizations, through religious dogma or in explorations of the inner universe. The transdisciplinary "Object" and its levels of reality, the transdisciplinary "Subject" and its levels of perception and the Third Party included, define the transdisciplinary model of the phenomenon.

The object, the subject and the *included third party* define the ternary structure of the relationship with water: Levels of reality of water, Levels of reality of the subject and Levels of reality of the *included third party.*

According to Nicolescu (2014), in the ternary structure of Realities in Table 3 below, various combinations can be made between the trios. In this way, you can better understand the various realities of an event, which involve the levels of reality of the object, the levels of reality of the subject and the *third party.*

The possibility of determining various levels of useful relationships in the analysis of concrete situations: water being the object of transdisciplinary research, analyzed at the various levels of realities, systemic complexity, and the subjectivity of the subject, combined in the various realities that present themselves.

These combinations confirm the complexity of the transdisciplinary experiment based on the inclusion of the *third party* and the levels of reality that surround them.

Table 3 - Ternary structure of different levels of reality.

Object	Subject	Third party included
Levels of organization	Structure levels	Integration levels
Levels of confusion	Language levels	Levels of interpretation
Physicist levels	Biological levels	Psychological levels
Levels of ignorance	Intelligence levels	Levels of contemplation
Levels of objectivity	Levels of subjectivity	Levels of complexity
Levels of knowledge	Levels of understanding	Levels of being
Materiality levels	Levels of spirituality	Levels of non-duality

Source: Nicolescu (2014, p. 61).

The transdisciplinarity of water lies in the various ways of recognizing, including and valuing the subject's relationship with water. This requires that the *third party included* be characterized as the main element of the trilogy - object/reality/complexity, manifesting itself in what seems contradictory, acting simultaneously as a dependent subject, an agent of environmental protection, a consumer/polluter or, in other roles, in the face of the various levels of reality, objective and subjective, presented.

Transdisciplinarity encourages critical and reflective action in the permanent process of use/care, from the perspective of social translation. It emphasizes the interrelationship and practical experiences of the subject, and the multiple social mediations. It brings not only new principles, but a different understanding of how to think and act in ecological practice.

Therefore, the purpose of analyzing water from a transdisciplinary perspective goes beyond disciplinarity, multidisciplinarity and interdisciplinarity. It goes beyond and through, whose study is associated with the dynamics of the multiplicity of dimensions of reality and classical scientific knowledge.

In this way, comparisons between the assumptions of scientific theories were analyzed within the exploratory methodology of the research using bibliography and literature with a qualitative approach. With the support of the transdisciplinary vision, especially through the articulation of the researcher with the object of study, in analyzing transdisciplinarity as a methodological option for broadening the reductionist vision of classical science on the subject of water, through the participation of the subject, listed as a *third party.*

In order to promote understanding of the guiding principles of the research, table 04 adapts the current environmental situation [water] to Santos' five principles, and highlights the specificities for broadening the reductionist and decontextualized view, given the diversity of the different levels of realities. Santos (2008) proposes five principles as aspects of teaching and learning processes to reduce the gap between the knowledge produced and its usefulness.

The transdisciplinary methodology for water presents a broad spectrum of interrelationships between the environment and living beings. The first principle determined by Santos (2008), the hologramatic principle, allows for many combinations between the various levels of reality of the object, the subject and the *third party included.* This systematic perception involves the principles of the complementarity of opposites, uncertainty and autopoiesis.

Table 4 - Adaptation of the five principles of Santos (2008), interpreted through the vision transdisciplinary water.

EDUCATION	WATER
Holographic: The parts can only be understood from the interrelationships in the dynamics of the whole, highlighting the multiplicity of interacting elements to the extent of the integration that reveals the existence of different levels of reality, opening up the possibility of new visions of the same reality.	Holographic: Water cannot be considered in a single context, but as *complementary in* the sense that only the totality of phenomena, or the complementarity of opposites, will show the possible information about it.
Transdisciplinary: It sets out to transcend classical logic, the logic of "yes" or "no", of "and" or "not and", which does not fit definitions such as "more or less" or "approximately", expressions that are "between dividing lines" and "beyond dividing lines", considering that there is a third term in which "and" joins "not and".	Transdisciplinary: The transdisciplinarity of water is based on the logic of the *third party included, making it* impossible to involve the subject without taking into account their own reality, involving all the complexity. It goes beyond simple use, because when you use water, you commit to it and take responsibility for it.
Complementarity of opposites: The dichotomy of binary pairs is opposed, referring to the level of integrated reality, i.e. reason "and" emotion, individual "and" society, health "and" disease, subjective "and" objective, *sapiens "and" demens,* good "and" evil, enclosure "and" openness of beliefs or theories. The conjunction "and" must be used to signify the joining, association and articulation of pairs.	Complementarity of opposites: Water is the natural element, characteristic of the complementarity of opposites, because it is dependent on the levels of reality, that is, it is merchandise "and" common good, sacred "and" popular, material "and" feeling, life "and" death, and health "and" illness.
Uncertainty: Manifested in the impossibility of ignoring subject/object interaction, i.e. separating the observer from nature, because the influence of	Uncertainty: Manifested in the relationship between water and the environment, and in the interaction between the observer and the water, which is the proper meaning of water, truth is relative.
observation and comparable to the observed phenomena.	
Autopoiesis: The living being is an autopoietic system, it self-organizes and self-constructs.	Autopoiesis: Water is the basic element for autopoiesis, necessary for the effective action of the process of self-organization; self-preservation and self-renewal, healing, homeostasis and adaptation in guaranteeing the life cycle.

Source: Adapted from Santos (2008).

Considering the transdisciplinary perspective, the axis of relations with other knowledge, comparative tables 02 and 04 help in the critical and reflective analysis of environmental phenomena and thus make it possible to perceive multiple realities and cross-references between methods and approaches.

The transdisciplinary experiment, determined by the triad of the third included in the combination of three terms: the level of reality of the Object, the level of reality of the Subject and the level of reality of the *third included, is* characterized by the simultaneity of action and the production of results. In the search for new knowledge and technologies for the care and protection of the hydrological cycle.

The research, based on the concept of the *included third party,* uses the technique of data analysis to collect the perceptions and knowledge of the social actors (subjects) and detect the different views of the realities that make up the research collective. The *included third party,* perceived through the subject's inability to dissociate themselves from water, through their own use and the product of this use, contributes to the construction and elaboration of knowledge at different levels of reality.

6 THIRD SEGMENT INCLUDED

Transdisciplinarity aims to stimulate an understanding of environmental complexity, to encourage the involvement of the subject and their multiple mediations with the environment. The research therefore involved ten subjects listed as *third parties.*

The sample consisted of nine citizens interviewed in two municipalities in the extreme south of Santa Catarina, Criciuma and Urussanga, characterized as: two *women;* one politician; three student workers; one environmental activist; one post-graduate student (master's degree) in Environmental Sciences; one business administrator. There was also the participation of a world representative, the director of NESTLE - foods, through an interview published on *youtube.com,* totaling ten subjects.

The women interviewed have a direct relationship with the research, through their daily activities with water. They are working mothers living in a condominium in Criciuma, who in 2014 suffered a period of drinking water shortages, and their supply was replaced by water from an artesian well polluted by coal mining.

The participation of women in environmental decisions regarding water was recommended by the 1992 Dublin Declaration in Ireland, because the role of women in the integrated management of water resources is relevant and a *key player* in the provision, management and protection of water. *Women* therefore play an important role in water, since historically they have been directly responsible, within the domestic context, for the hygiene, health and nutrition of children and the family.

Women participate as the *third party included,* because their cultural and historical reality is taken into account, since in their daily dealings with water they maintain a differentiated and unique relationship. They are characterized as agents with the closest relationship to water and holders of useful knowledge in valuing it, making decisions and taking care of it. In their effective participation, women collaborate with their own knowledge and know-how about the relationship between water and the life cycle, empowering them as agents of protection.

The participants' perceptions of water are based on the need to protect their families and their daily chores, their own survival and the need to take care of their own health and food. However, despite having suffered the consequences of the lack of water and the use of water that is unsuitable for consumption, the participants' perceptions are fragmented, and the women interviewed do not recognize the complexity of the relationship with water and the environmental risks in the region that compromise the quality and quantity of water provided.

Interviewee B talks about the water conditions during the period in which the condominium where she lives had its drinking water supply interrupted and was replaced by water from an artesian well, made in degraded soil and mined by coal. "Before, we couldn't drink the water, the children would vomit, their hair would fall out, all because of that water, it tasted bad." (Interviewee B).

The interviewee's reality, the complexity of the social phenomenon and the importance of women's contribution to environmental situations, specific to water, can be seen in her account. The reality that involves family health and social well-being is understood through the woman's daily life. Interviewee B, when discussing coal mining in the region and water pollution, said that she didn't think that mining could compromise the water: "I don't think so, do you? Not coal, right? Because coal comes from sulphur, just like the water we were drinking there, it was pure sulphur."

The perception is fragmented and the level of reality is immediate in understanding water as life, without realizing the relationship between the conflict and the reality exposed and the history of coal mining in the region. In this way, the participation of these women is characterized by the need to know, to know how to effectively participate in environmental actions, relating domestic knowledge to environmental knowledge in the elaboration of knowledge and, finally, to become an agent of protection.

Water's long journey through the hydrological cycle is difficult for urban citizens to understand, since life in cities and the large-scale collection and distribution of both water and food has distanced human beings from nature, making it difficult to understand how life depends on the environment. The urban processes of collection and distribution, carried out by sanitation companies, have fragmented knowledge and recognition of the water cycle.

Citizens are not aware of the natural process of the renewal cycle and are not committed to protecting water. Understanding this process is difficult because it is not present in everyday life, since drinking water appears in pipes and spouts, in quantity and quality for daily needs. Without the subject's effective participation in environmental actions, a holographic understanding of water is not possible. The decontextualization of the hydrological cycle, with little or no relation to the environment and human action that pollutes bodies of water, disengages them from the whole process.

The responsibility for care falls to the distribution companies or the state, the citizen does not feel "part" of the process and the participation of women as an *included third party* contributes substantially to this recognition. In fact, by taking on roles in the triad of the "included third party", it is possible to collaborate with women's own knowledge of everyday life. When she takes on the role of "object" in the triad, in her interaction with water, she has the possibility of developing new knowledge about her worldview and the social relationships she has with the environment. Finally, when she is listed as a "citizen", she assumes attitudes towards caring for this resource, which is so valuable to life and the family.

The need for *women*'s participation and integration as contributors to the diverse knowledge developed through daily experience with the family and children is also formative for their own understanding of the [global] environmental complexity surrounding water.

Unlike interviewee C, a participant in the category of environmental defense agent, involved in the protection of the Urussanga river basin, his perception is amplified by his own experience with environmental causes, the soil and the waters of the region, as he says:

> In reality, water is always the same on the planet. The same water you have today. It comes and goes. It's just that whoever occupies the space on the planet transforms the availability of that water, in quantity and quality, depending on human intervention. [...] Without water there is no life and there is no economy!" (Interviewee C).

The speech denotes the holographic perception of the close and direct relationship between water and life, when he emphasizes: "[...] all segments, all sectors must move towards preserving the quality and quantity available, for current and future [living beings]". (Interviewee C).

Interviewee C understands that water has different levels of reality and mentions some of them, such as territorial reality, the use of living beings and the ecosystem itself. He also relates water as a factor in raising awareness of the problem so that everyone can participate. The distribution of water, with quality and availability for living beings.

To determine the interviewees' perception of the principle of the *complementarity of opposites, they* were asked how environmental problems and social conflicts can contribute in some way to water protection:

> [...] the lack of water, to meet this need, is growing every day and more people on the planet lack drinking water, so the good thing about this problem is that it's starting to awaken in people the need to participate in the search for a solution (Interviewee C).

Faced with the logic of the contradictory, many aspects are considered incomprehensible and even absurd, since they are analyzed in the classic logic of being or not being, but the contradictory aspects are no longer absurd in a logic founded on the postulate "and this and that", or rather, "neither this nor that",

The participation of the subject is understood by interviewee C, who analyzes this participation in the face of conflicts with water, its availability and quality for all, and the subject as an agent in the search for solutions. Thus, the *updating of* interviewee C's knowledge of the phenomenon of water, aggravated by the history of neglect and lack of investment in sustainable actions, has the counterpart - according to Lupasco's 'contradiction' - of *enhancing* his own participation in projects and actions, in the development of attitudes to protect water.

Interviewee C also says that there is a need for participation, adequate investment and public actions with public policies to solve the current problems.

> Water management in Brazil is participatory, with the participation of all segments, water users, civil society and the government sector. Because it has to have everyone's vision and everyone looking at water as a common good and that the actions and form of management

have to have everyone's participation (Interviewee C).

In view of the perception expressed by interviewee C, the holographic vision is notorious, when he explains the understanding of the environmental scope of water, and says: "Because it has to have everyone's vision and everyone looking at water as a common good and that the actions and form of management have to have everyone's participation."

Another transdisciplinary characteristic of interviewee C is his understanding of the levels of reality, to the extent that he understands that, "[...] without water there is no life and there is no economy! "

Participation as a subject in decisions and discussions makes it possible to see beyond the initial viewpoint, in the face of the various perspectives that perceive it as a vital asset, a commodity, a sacred element, among others, as well as practicing citizenship and tolerance - ethical aspects that cut across education.

Two university students and workers also took part in the research and the young people's perception of the world brings us to another level of reality, often devoid of preconceptions, thus bringing renewal and the perception that the new is possible.

Young people participate as "citizens", contributing ideas and suggestions for use and care; as "subjects" of the experiment, as they learn and build environmental knowledge in a participatory way; and as *third parties,* with the level of reality perceived through the original and innovative vision of youthful thinking.

The young man interviewed, E, realizes that when it comes to the problem of water: "There are several ways to solve it and there are also several solutions. If degradation stops, the environment can be regenerated." The speech quoted above can be analyzed in the light of the problem statement, when it opens up the possibility of other options for solving a problem. It reaffirms the proposition of relative truth, of Santos' principle of Uncertainty, based on the interrelationship between the subject and the object.

Interviewee E seems to relate the theme to the different levels of reality and the various possible solutions, as well as the notion of autopoiesis, a term proposed by Maturana, when he mentions that the environment has the capacity to regenerate.

With original and innovative ideas and the ease with which they deal with technology and new information, young people have a greater perception of their interrelationships with the world. The participation of young people in everyday actions promotes the critical exercise of autonomy and citizenship, developing ecological awareness and ensuring sustainable future actions.

The ease with which young people can understand the relationship between global causality, ecological reasoning and responsibility, and their own actions, places them in the position of an *included third party*.

Another personality in the research who is important in the management of water resources is the politician, represented here by a councillor whose perception is the great challenge related to sustainable development, the guarantee of education and health, in the promotion of life and society. Interviewee D, the politician, said:

> [...] people are very focused on development and actually forget that it (development) has to go hand in hand with sustainability. Yes, you have to develop! And it has to have an impact on the environment and move towards development with as few impacts as possible. (Interviewee D).

The political personality represents the "subject", the agent of water resource management. The perception of public managers must be amplified by ecological and historical-cultural knowledge of the region, as a guarantee of water management beyond personal and territorial boundaries. In public management, academic training becomes the differential, raising the level of commitment and political decisions in the face of environmental problems.

Another reality exposed is that of water as a commodity, which in periods of crisis and scarcity acquires economic value and the market becomes attractive. In this context, the research also included the perception of the food industry, represented by NESTLE foods, the world's largest bottler of drinking water. This participation was made possible through an interview published on www.youtube.com, in which the director and president Peter Brabeck-Letmathe expresses his opinion on two issues relating to water, one as a right and the other as a food that should be priced.

In the view of Peter Brabeck-Letmathe, chairman of the Nestle group, water should be considered a food like any other and should be priced. He believes in the need to privatize the supply of water, as a way of making consumers aware of its importance, because the lack of a "price" leads to underemployment and loss of value.

According to the president, Brabeck, water will be valued and those who have access to water can contribute to those who do not:

> I personally believe that it is better to price a food so that we are all aware of its price and then take specific measures for the part of the population that has no access to this water at all, thus having many different possibilities (BRABECK, 2013).

The contribution of the market perception becomes important, since water is considered food and is a reality. In turn, those who advocate the privatization of water or the privatization of water resources use the argument that water must be cared for, preserved, captured, treated and distributed, and that this generates costs that the state, through public policies, is sometimes unable to manage. In this scenario, national and multinational companies come in, arguing and transforming water into a commodity.

The president of NESTLE, in his speech, reaffirms the problem of this research, which is configured in water, as an element of discussion in the context of environmental complexity, because he defends the idea of water as any food.

The larger discussion of water in the environmental context is based on two strands of thought: the first is sustained by the idea of water as a common good for humanity and in need of care, and the second is defended as a commodity, incorporated into the logic of the market, of supply and demand.

Brabeck's words have provoked opposition from internet users when he proposes a project to privatize water and promote "sustainability". They accuse him of condoning slave labor and the exploitation of water bodies for capitalist ends, especially when he suggests that the perception of free water makes people disregard it and waste it.

Brabeck, in his narrow view of the market, fails to notice the holographic character of water, perceiving it as just *any food, and* proposes the dominance of those who will profit from the trade over those who do not have access to drinking water, when he says: "[...] and then let's take specific measures for the population that has no access to this water at all, so there are many possibilities out there".

The capitalist perception does not value the need for popular participation in decisions about water. Access to water becomes passive and dependent on trade. It places the right to water in the background and in dependence on the market, subjecting it to the laws of the market and to the owners of that market. Therefore, "Transdisciplinary ethics rejects any activity that avoids dialogue and discussion, whatever its origin - ideological, scientific, religious, economic, political or philosophical." (NICOLESCU, 2000, p.7).

Pricing water, transforming it into an economic resource, can increase environmental conflicts, which have their roots in the distribution of quality and quantity, excluding and increasing the distance between peoples and regions without water and basic sanitation. It can also, on the contrary, in the perception of the complementarity of opposites, generate greater financial resources and development for regions that have an increased demand for water, not just for cumulative purposes, and thus allocate part of the resources to the management of water sources in a sustainable way.

In this way, promoting water through and beyond supply and demand increases the radius of the capitalist vision, with the critical analysis and recognition that the attitude of pricing water has the possibility of increasing the social distance that the supply of water has been promoting between peoples.

In the same way, interviewee J, representing the businessmen of the coal region, exposes his own reality when he mentions his life history and the symbolic relationship between water and life, but he does not perceive the interdimensional subjective reality of water with the cycle of life and "interlife".

The problem of water is reported in the face of physical reality, immediate and present. The interviewee analyzes the realities, supported by the classical paradigm and exposes his own thinking, determining the diversity of realities on a single level, that of availability. So when asked. About the relationship between the cycles of *water and life,* he *says:* "I've never thought about it specifically. "

Without valuing the subjectivity of water with the various levels of realities, such as religious symbolism, cultural value and sentimental value for some peoples. He does not establish the relationship between the water cycle and the cycle of life, citing that in religion, *"water is the beginning of life"*. From the holographic analysis, we can determine the inseparability of life and water. The relationship between the two cycles goes beyond the polynomials of the hydrological cycle: *evaporation-condensation-precipitation-infiltration* and the cycle of life: ***fertilization-birth-life-death.***

It can be thought that the trajectory of life, past and future, lies in the continuous interaction between the cycles, determined by the combination of the levels of reality of water, the subject and the third party. In view of the above, Interviewee J's sentence shows the holographic perception, when he says: "[...] in the renewal of the water cycle, I think the same thing happens with the life cycle, except that if you treat it separately, water will always renew itself and life, it will deteriorate." (Interviewee J).

Interviewee J refers to the "different realities", but on the physical and immediate level, without realizing the "levels" of reality of the various dimensions presented. When asked about his own perception of strictly physical reality and whether he could transcend it, he states:

> I only think about the question of the availability of use, [...], that depending on the degree of scarcity, how indispensable it is for life. Scarcity promotes creative and differentiated alternatives according to the realities and where it is abundant, perhaps the culture of scarcity will never pass, how to survive without it. So they are different realities according to availability (Interviewee J).

However, due to Interviewee J's education, level of understanding and capacity for reasoning, it can be seen that if he is asked to respond to these subjective realities, he is able to make a critical analysis of the issue and relate the symbolic and sentimental value of water to cultures, religions and other knowledge, given their respective values. However, the symbolic, sentimental and religious valuation of water might not be considered equitable in contributing to the solution of socio-environmental conflicts, given the difficulty of perceiving the levels of realities and the "real" value, even if subjective, of water.

The difficulty of subjective valuation emerges in the reports when interviewee J says: "How indispensable it is for life". As if the indispensability of water for life were subject to levels or degrees of scarcity, as if the need for water for life could be graded and/or questioned. However, when analyzed in terms of the interrelationship between life and water, water is essential.

The interdimensionality represented by the historical subjectivity of the manifestation of living beings through water in human evolution and its own future projection, beyond the materialist and immediate paradigm, is difficult to understand and is perhaps the greatest obstacle to transdisciplinary perception.

Thus, when referring to the transdisciplinarity of water, it is necessary to perceive this natural element as a phenomenon[4] , which has been transformed over time and currently, in view of the perspective of life, basically depends on it.

Finally, the perception and participation of those involved in water issues is based on the theory of the *included third party,* because it goes beyond and through reality itself. Not only as a contributor of their own knowledge, but also as a beneficiary of the process, whether in providing for the future, understanding the challenges, producing and elaborating more complex knowledge, or promoting and improving quality of life.

[4] PHENOMENON. According to Emmanuel Kant, it constitutes the world as we experience it, as opposed to the world as it exists independently of experience.

7 FINAL CONSIDERATIONS

The environmental liabilities inherited in the region, the constant vertical growth of cities and urban densification, compromise the citizen with the consequences of their own manifestation in the search for quality of life, through socio-environmental solutions. Transdisciplinarity stimulates the subject's critical and active participation, pointing to "a new path" in the permanent process of [water] use, as an important step towards the inclusion of the "third party".

Despite being essential for life, caring for water is still interpreted only as controlling waste. The difficulty in understanding the meaning of *care,* i.e. how to deal with water responsibly and sustainably, is noticeable. Water needs *care,* much more than rationing, so the knowledge produced through the subject's experiences, shared in environmental actions and decisions, is the key to the process of appropriating water.

Complex measures involve care, because water is the main focus of the socio-environmental relationship, and after use, it must be properly returned and integrated into nature. Environmental education teaches care and reinforces the need to change the population's daily habits, with attitudes to save water, making the subject responsible, but the basis of the problem lies in the lack of water protection.

The indiscriminate use of natural resources, perceived as unlimited, consumes and wears down the environment, affecting the autopoiesis of systems. In order to grasp the complexity of this fact, it is necessary to perceive the various "realities" in a juxtaposed, diffuse and simultaneous way presented in the interaction between living beings and water. Otherwise, one is susceptible to reductionism, when one perceives only the physical, immediate and unilateral reality.

The current situation of scarcity in some regions of the country requires changes in the way we perceive water. The appropriation of water goes beyond collection, distribution and ethical use, perceiving it as a necessary asset for any human manifestation, to maintain life or produce quality living conditions.

To paraphrase Gerald Holton (1984, p.52), when you ask: What is water? The answer is: the consumer, the various types of handling, the experiments used, the knowledge produced, the theories and models of interpretation, the environment and everything else that complements the context, otherwise there would be no water. All this together is water.

Accepting transdisciplinarity as a methodological attitude, in the form of empathy and openness to others and their knowledge, enriches research and raises the quality of studies, since it contributes to the reliability of the results.

Transdisciplinarity reveals the relative or approximate truth, because it is not possible to anticipate or determine the limits of reality, only experience can indicate them. Determining new possibilities for understanding the different perspectives and meanings of the importance of water in the different cultures of the world, which, in addition to mere technical knowledge, makes it possible to involve *third parties in* order to value participation. When the other person feels part of the process, the guarantee of commitment and change is more evident.

Understanding the multiple mediations and the relationship with the environment characterizes the citizen as the *included third party,* acting in the various roles of interaction, sometimes as a consumer, sometimes as a polluter, sometimes as an agent of water protection.

Activities and decisions related to water, in a *transdisciplinary* way, are capable of highlighting the subject as a participatory agent, developing and changing attitudes for better use and aiming to guarantee equal access to drinking water, including for those who have not yet been born.

Transdisciplinarity is seen as a tool for educating and informing citizens, committing them to politics and current socio-environmental issues that tend to affect life on Earth. It makes it possible to increase commitment to water as you use it, because the more you use it, the more you take responsibility, preserve and conserve it, in the context of your own reality and the present world, reversing the actions for your own benefit.

It is possible to innovate and think differently by looking at reality from a new point of view. Innovative ideas are initially viewed with disbelief, due to an inability to perceive the new outside the parameters of the traditional reductionist view. In order to innovate, it is necessary to understand whether the individual is part of the majority, which reflects the need for change.

One of the pillars of transdisciplinary research, the *third one included*, whose quantum argument is that it is impossible for the observer not to influence the object of the research and their own exemption in the process as an agent of innovation, the researcher moves from the position of researcher to that of participant, because the research influences the researcher, who influences the research and the environment, and the environment consequently influences both.

Also recorded as a result of the research is the report of the *third party included,* that is, the speech of the researcher, participant and member of this experiment, who influenced and was influenced by it. The research made it possible to understand the transdisciplinary presuppositions in the construction of knowledge and in the elaboration of Environmental Science knowledge and in the manifestations as a citizen, in the face of the diverse realities of water, which the senses and understanding allow, contributing more and more to the sustainability of life.

The characteristic of transdisciplinary research is that it is diffuse, simultaneous and spiral. It is confirmed when it results in the production of knowledge capable of contributing to conflict resolution, as well as improving the critical capacity of the subject, the *third party included,* culminating in the promotion of life.

8 REFERENCES

AGUDO, Marcela de Moraes; TEIXEIRA, Lucas Andre. **Water as a commodity.** Bauru City Newspaper. Digital edition. 05.11.2014. Available at: http://www.jcnet.com.br/editorias noticias.php?codigo=236396. Accessed on January 15, 2015. ' - - -

AMORIM, Joao Alberto Alves. **Water Law:** The Legal Regime of Freshwater in International and Brazilian Law. Sao Paulo: Lex, 2009. 368 p.

BACHELARD, Gaston. **Water and Dreams:** Essays on the Imagination of Matter. Original title *L'eau et lesreves.* Translated by Antonio de Padua Danesi. Antonio de Padua Danesi. Rev. Rosemary Costhek Albino. Coord. Mauricio Balthazar Leal. Sao Paulo: Martins Fontes, 1998. 202 p.

BOUGUERRA, Mohamed Larbi. **The Battles of Water:** For a Common Good of Humanity. Original title in French: Lesbatailles de l'eau - Pour um biencommun de l'humanite, Petropolis: Vozes, 2004. 204 p.

BRAZIL, **Law No. 9.433 of January 8, 1997**: Institutes the Water Resources Policy, creates the National Water Resources Management System, regulates item XIX of art. 21 of the Federal Constitution, and amends art. 1 of Law No. 8.001 of March 13, 1990, which amended Law No. 7.990 of December 28, 1989.

, ANA, National Water Agency. **Caring for water:** solutions for improving the quality of water resources / National Water Agency; United Nations Environment Program. Brasilia: ANA, 2011. 160 p.

, ANA, National Water Agency. **Water resources in Brazil.** Brasilia: ANA, 2013. 434 p.

CAPRA, Fritijof. **The Tipping Point.** 10 ed. Sao Paulo: Cultrix 1995. 448p.

The tao of physics: A parallel between modern physics and Eastern mysticism. Sao Paulo: Cultrix. 1983. 260 p.

CASAN. **Annual report on the quality of water distributed.** SAA of Criciuma agency. 2014. Legal representative: CEO Valter Jose Gallina. Available at http://www.casan.com.br/ckfinder/userfiles/files/rel anu qual agua 2014/srs/criciuma/criciuma.pdf#4098Accessed 11 Jun. 2015.

CASTRO, Jose Esteban. **The privatization of water and sanitation services in Latin America.** Revista NUEVA SOCIEDAD No 207, January-February 2007, ISSN: 0251-3552.
Available at www.nuso.org. Accessed on 10 Dec. 2014.

CET-Water. **Water and Transdisciplinarity Charter.** Summary document of the 1st International Seminar on Water and Transdisciplinarity - towards an ecology of knowledge. Brasilia, 2011. 2p. Available at:http://cetagua.org/carta-de-principios-centro-de-estudo-transdisciplinar-da-aguahttp://cetagua.org/carta-de-principios-centro-de-estudo-transdisciplinar-da-agua-cet-agua/cet-agua/.Acesso on 28.10.2012.

DOMINGUES, Ivan. **Knowledge and transdisciplinarity II:** Aspects Methodology. Ivan Domingues (org.). Belo Horizonte: FMG, 2005. 424 p.

FLORES, Karen Muller. **The Recognition of Water as a Fundamental Right and its Implications.** Rio de Janeiro: RFD- Revista da Faculdade de Direito da UERJ, v.1, n. 19, jun./dez. 2011.
Available at:
http://www.google.com.br/url?sa=t&rct=i&q=&esrc=s&source=web&cd=1&ved=0CB4QFjAA& url=http%3A%2F%2Fwww.e- *
publicacoes.ueri.br%2Findex.php%2Frfdueri%2Farticle%2Fdownload%2F1724%2F1337&ei=X2 ' * * '
4FVbrTMYHEgwSh0YEo&usg=AFQiCNHaO6rdPUJoaP2ElauMmzqXaOEMOg&sig2=5M7Hth4bNOi989tzGWscAA&bvm=bv.88198703,d.eXY. Accessed on June 14, 2014.

FREITAS, Juarez. **Sustentabilidade:** Direito ao futuro. 2 ed. Belo Horizonte: Forum. 2012. 348 p.

HOLTON, Gerald. **Principle of complementarity**. Dinorah de Oliveira Mendes (transl.). (Vol.
II, n°09, Oct/Dec, 1984 p.49 - Humanity). Available at:
http://www.fflch.usp.br/df/opessoa/Holton-Complementaridade.pdf Accessed on July 10, 2014. * * * *

LEFF, Enrique. **Sustainable Discourses.** Mexico: Siglo XXI. 2008. 272 p.

MATURANA R., Humberto **Cognition, science and everyday life**; organized and translated by Cristina Magro and Victor Paredes. - Belo Horizonte, UFMG, 2001. 203 p.

MONTANARI, Valdir; STRAZZACAPPA, Cristina. **Along the paths of water.** Sao Paulo: Moderna, 1999. 56 p.

Humberto R.; VARELA, Francisco J. **The tree of knowledge:** the biological bases of human understanding. 6. ed Sao Paulo: Editorial PsyII, 1995. 283p.

MORAES, Maria Candida, **Interdisciplinarity and transdisciplinarity in education:** ontological and epistemological foundations, problems and practices in: **Transdisciplinarity and ecoformation:** A new look at education. Original title Transdisciplinariedad y Ecoformacion: Una nueva mirada sobre la educacion. Sao

Paulo, TRIOM, 2008. p. 61-86 432 p.

MORIN, Edgar. **Introduction to complex thinking.** Trad. Eliane Lisboa. 4 ed. Porto Alegre: Sulina. 2011. 120 p.

NICOLESCU, Basarab. Contradiction**, Logic of the Third Included and Levels of Reality,** Paris, paper presented at Ateliers sur la contradiction, Ecole n.s. des mines, 2009.

Basarab; BADESCU, Horia. **SthephaneLupasco:** the man and the work. Original title *SthephaneLupasco.* Sao Paulo, Triom, 2001. 344 p.

Basarab. **Education and transdisciplinarity.** Brasilia, UNESCO, 2000.

Basarab. **The Manifesto of Transdisciplinarity.** Sao Paulo: Triom, 1999.

Transdisciplinarity and university education: emerging theories and practices. Mexico: CEUArkos, 2014.

Basarab, Lecture at the International Congress **The University's Responsibility to Society**, International Association of Universities, Chulalongkorn University, Bangkok, Thailand, November 12-14, 1997.

PALANIAPPAN, MEENA; et al; **Caring for Water:** Solutions for improving the quality of water resources. Brasilia, ANA, 2011. Available at: http: //www.pnuma.org.br/admin/publicacoes/texto/Cuidando_das_aguas_final_baixa.pdf . Accessed on 15.02.2014.

PETRELLA, Riccardo. The **Water Manifesto: Arguments** for a world contract. Original title - The water manifesto - Arguments for a world contract. Trad. Vera Lucia Mello Joscelyne. Petropolis, Vozes, 2002. 160 p.

PHILIPPI Jr, Arlindo et al. **Interdisciplinarity in environmental sciences.** Sao Paulo, Signus, 2000.

SACHS, Ignacy. **Towards Ecosocioeconomics:** Theory and Practice of Development. Org. Paulo Freire Vieira. Sao Paulo, Cortez, 2007.

SANTOS, Akiko. Complexity and transdisciplinarity in education**:** five principles to recover the missing link. **Revista Brasileira de Educagao,** Rio de Janeiro, ANPED, n.37, (apr. 2008), p.71-83 ISSN: 1413-2478.

TEIXEIRA, Wilson. **Deciphering the Earth.** Sao Paulo, Oficina de textos, 2001. 557p.

THEODORO, Suzi Huff. Conflicts and sustainable use of natural resources, in: **Considerações sobre o conflito e uso sustentavel em recursos hidricos**. Sao Paulo, Triom, 2008.

RSROCKNROLLER. Interview with Peter Brabeck. **Water is not a basic human right.** Published on 29.04.2013. Available at:https://www.youtube.com/watch?v=5a8qzsM9Kqg . Accessed on January 15, 2015.

MIX
Papier aus verantwortungsvollen Quellen
Paper from responsible sources
FSC® C105338

Printed by Books on Demand GmbH, Norderstedt / Germany